Puzzle It!
Attribute and Pattern Puzzles

Written, Designed, and Illustrated
by Kathleen Bullock

Incentive Publications
Nashville, Tennessee

Cover by Angela Stiff
Edited by Marjorie Frank and Jill Norris
Copy edited by K. Noel Freitas and Scott Norris

ISBN 978-0-86530-519-9

Copyright ©2008 by Incentive Publications, Inc., Nashville, TN. All rights reserved. No part of this publication may be reproduced, stored in a retrieval system, or transmitted in any form or by any means (electronic, mechanical, photocopying, recording, or otherwise) without written permission from Incentive Publications, Inc., with the exception below.

Pages labeled with the statement **©2008 by Incentive Publications, Inc., Nashville, TN** are intended for reproduction. Permission is hereby granted to the purchaser of one copy of **PUZZLE IT! ATTRIBUTE AND PATTERN PUZZLES** to reproduce these pages in sufficient quantities for meeting the purchaser's own classroom needs only.

1 2 3 4 5 6 7 8 9 10   11   10   09   08

PRINTED IN THE UNITED STATES OF AMERICA
www.incentivepublications.com

Contents

Welcome 5
How To Use 6
- #1   Kick-Flipping Attributes 7
- #2   Swing Your Partner 8
- #3   Snappy Beads 9
- #4   Space Woogies 10
- #5   Puzzling Discs 11
- #6   Note-ables 12
- #7   Quick Thinks 13
- #8   Making Matches 14
- #9   Carousel Comparisons 15
- #10 Picture This! 16
- #11 Mug Shots 17
- #12 Does It Fit? 18
- #13 Confusing Confections 19
- #14 Going Buggy! 20
- #15 Puzzling Patterns # 1 21
- #16 Seeing Dots 22
- #17 Off to the Races 23
- #18 Puzzling Patterns # 2 24
- #19 The Almost Leaning Tower 25
- #20 Double Take 26
- #21 Who Stole the Dough? 27
- #22 Riddle Me an Attribute 28
- #23 Spy Talk 29
- #24 Talking Patterns 30

#	Title	Page
#25	Fly Trap	31
#26	The Eye of the Beholder	32
#27	Mirrored Patterns	33
#28	Logo Logistics	34
#29	Germ-inations	35
#30	Attributes with "Class"	36
#31	Overlapping Pattern Chains	37
#32	Points of Style	38
#33	Interesting Choices	39
#34	Number Puzzlers	40
#35	The Catch of the Day	41
#36	Puzzling Patterns # 3	42
#37	Stuffed Lockers	43
#38	Cross-Number Puzzle	44
#39	Life in a Fish Bowl	45
#40	Toss the Dice	46
#41	Rah-Rah Combos	47
#42	Puzzling Patterns # 4	48
#43	Mini-Mysteries	49
#44	Analogy Quick Thinks # 2	50
#45	Pizza Portions	51
#46	Prizewinning Numbers	52
#47	Predictable Patterns	53
#48	Cipher Words	54
#49	Book Bewilderment	55
#50	A Tribute to Flags	56
#51	Squared!	57
#52	Eighth Grade Musical!	58
Answer Key		59

WELCOME

To The Fun of Attribute and Pattern Puzzles

Everybody loves a puzzle! A puzzle is like an unsolved mystery, teasing you to be the one that unravels it. There are few things that match the feeling of satisfaction you experience when, after thinking long and hard about a puzzle, the solution suddenly materializes—clear as crystal. It is a truly satisfying moment.

Try this puzzle magic: when students wrestle with a puzzle, classroom learning is energized. Who can walk away from the invitation to tackle a puzzle? It's too much fun to try to figure it out. Even the most reluctant students seem to wake up and be drawn into the solution process.

But puzzles are much more than just fun! They give the brain a workout and nurture cognitive processes. Every classroom and home should offer many puzzling opportunities. The critical thinking and problem-solving skills that are honed apply to every facet and subject area of learning. In solving puzzles, students make use of higher level thinking skills such as logic, analysis, induction, deduction, synthesis, sequencing, and creativity—and they often must use several of these simultaneously. They must observe, ask questions, consider strategies, try different strategies, visualize different possibilities, think "outside the box," and figure out why one thing works and another does not. Many puzzles also refine hand-eye or hand-mind coordination, spatial awareness, and mental gymnastics.

Puzzles must be a part of every serious curriculum. The puzzles in the **Puzzle It!** series challenge students to analyze information and use their brains!

About The Puzzles In This Book

Attribute and Pattern Puzzles offers a rich array of puzzles that require puzzle-solvers to look for all kinds of characteristics in all kinds of arrangements. The eye and mind are challenged by minute differences and unexpected combinations. Each of these puzzles ignites the puzzler with delightful surprises and tricks as they seek out attributes and patterns lurking in shapes, numbers, words, and pictures. Examining attributes and patterns will significantly advance math skills, but these puzzles are not only for the math classroom. As puzzlers decipher the patterns or separate out the attributes that matter to the puzzle, they sharpen their observation abilities and strengthen skills that apply to all areas of the curriculum.

How To Use The Puzzles

- Look over each puzzle carefully. Read the instructions a few times.

- Consider the puzzle thoughtfully. Make sure the purpose is clear to you.

- Focus in on the attributes that are important in answering the questions or deciphering the pattern.

- Experiment with different strategies and different ideas. Try out different solutions.

- Take one puzzle at a time. A puzzle will grab you and won't let go until you figure it out. So let it swirl around in your head. Stay with it until you reach a solution.

- Check the solution against the original instructions or questions. Make sure the solution really works!

- Try not to peek at the answers. Ask someone else for an idea or a hint.

- You can tackle a puzzle alone, or share a puzzle with one or more friends, and tackle it together. Share ideas, discuss, argue—until you arrive at a solution.

- Remember that any given picture, design, person, geometric figure, word, or phrase generally has many attributes. In these puzzles, be sure to separate out only the attributes that apply to answer the question.

- When you find a solution, discuss it with someone else. Explain the steps and strategies you used to reach your answer. Compare your solution and methods with someone else's.

- There is something mesmerizing about attribute puzzles and pattern puzzles. You will want to create some yourself. Don't be afraid to try some! Once you get the idea of the different ways to form these kinds of puzzles, follow some of the same patterns, and write or draw puzzles to pass on to your friends and classmates.

About The Solutions . . .

The answer key gives solutions for all of the puzzles. In some cases, however, there may be more than one solution that makes sense. Give yourself or your students credit for any solution that can be explained and justified.

Kick-Flipping Attributes

The members of the Lightning Bolt Skateboarding Club decided to have all their skateboards painted with a common attribute. Some of the skateboards below belong to the Lightning Bolts, and some do not.

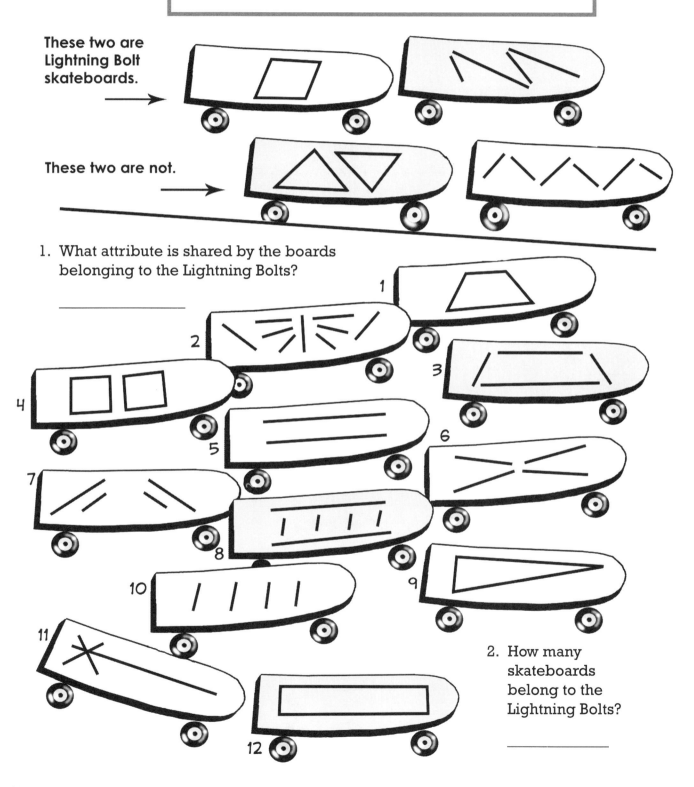

These two are Lightning Bolt skateboards.

These two are not.

1. What attribute is shared by the boards belonging to the Lightning Bolts?

   _____

2. How many skateboards belong to the Lightning Bolts?

   _____

Name _____

Puzzle It! Attribute and Pattern Puzzles © 2008 by Incentive Publications, Inc., Nashville, TN.

Puzzle 2

SWING YOUR PARTNER

It's the night of the Harvest Moon Square Dance.
Two couples have found their partners.
Name two attributes that each of the couples share.

_____

_____

Draw lines so that all the dancers have a partner. All the couples must share the two attributes you have named.

THESE ARE PARTNERS.

Name _____

Puzzle It! Attribute and Pattern Puzzles   8   ©2008 by Incentive Publications, Inc., Nashville, TN.

Puzzle 3

SNAPPY BEADS

Study the beads on the half-finished necklace. The beads represent two repetitions of the pattern the necklace maker will follow. Describe the pattern.

_____
_____
_____
_____
_____

Continue the pattern by choosing the next six beads from the box and drawing them on the string.

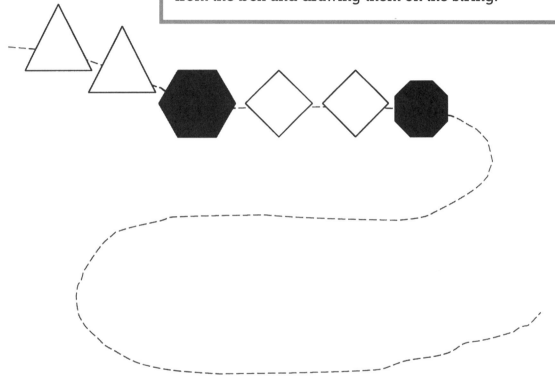

Name _____

Puzzle 5: Puzzling Discs

These titles are in Esteban's CD collection:

These titles are not in Esteban's collection:

Write a title that would fit into Esteban's collection: _____

These titles are in Sophia's CD collection:

These titles are not in Sophia's collection:

Write a title that would fit into Sophia's collection: _____

These titles are in Devon's CD collection:

These titles are not in Devon's collection:

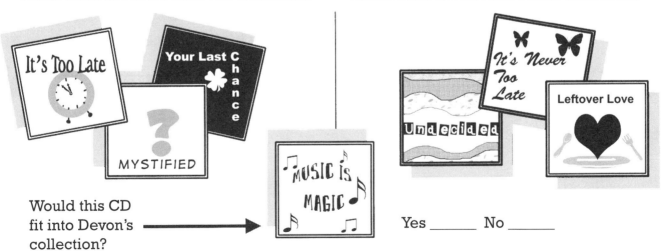

Would this CD fit into Devon's collection? →

Yes _____ No _____

Name _____

Puzzle It! Attribute and Pattern Puzzles   11   ©2008 by Incentive Publications, Inc., Nashville, TN.

Puzzle 6
NOTE-ABLES

The attributes of written music are fun to discover. Look for patterns in the examples below and complete each musical theme.

1. Blackbirds on wires make a pattern. The pattern repeats when new birds join the group. Draw them.

2. Study the pattern of the notes in section **a**. Use it as a guide to complete the pattern in section **b**.

3. Look carefully at the patterns in sections **a.** and **b.** Finish the pattern started in section **b**.

4. Create your own repeating pattern.

Name_____

Puzzle 7

QUICK THINKS

1. Finish the analogy.

2. Finish the analogy.

3. Choose the right answer from the boxes at right. Circle it.

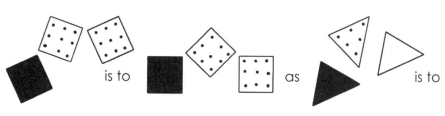

4. Finish the analogy.

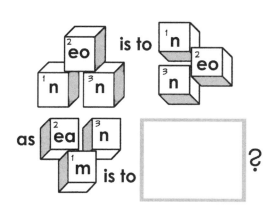

5. Finish the analogy.

Name _____

Puzzle It! Attribute and Pattern Puzzles   ©2008 by Incentive Publications, Inc., Nashville, TN.

Making Matches

Separate the pictures into four groups each with a common attribute. All the picture boxes have letters in the corners. Write the letters from each group on the lines below.

Unscramble each set of letters to form a word that names the attribute of that group. Write the words in the boxes.

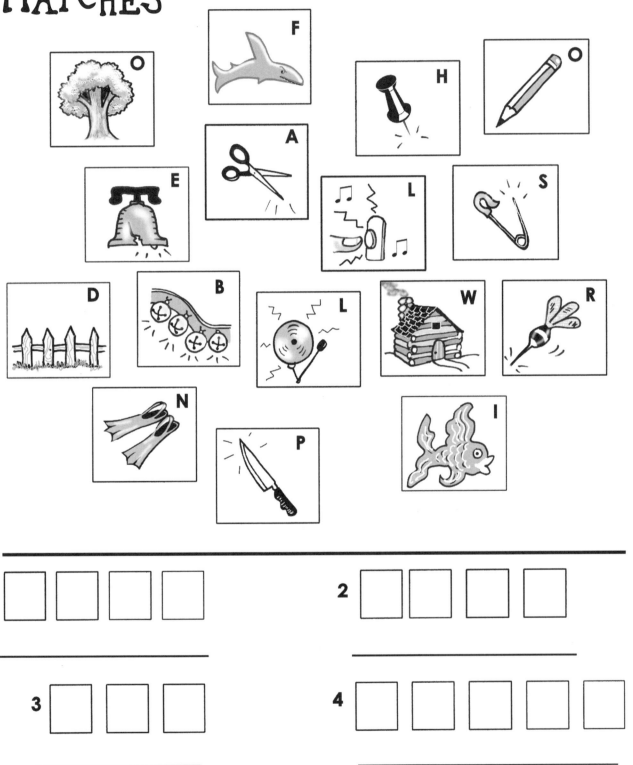

1 ☐☐☐☐

2 ☐☐☐☐

3 ☐☐☐

4 ☐☐☐☐☐

Name_____

Carousel Comparisons

Karina Karver is a famous carousel horse sculptress. Her new models are just a little different from the older ones on the carousel. Find the new horses Karina sculpted and circle their numbers.

Puzzle It! Attribute and Pattern Puzzles

Picture This!

Analogies are solved by discovering relationships between the things being compared. Finish each analogy by choosing the best picture in the frame to complete it. Draw your answers in the boxes or on the lines provided.

1. Food is to body as gas is to ☐

2. 289 is to 17 as 121 is to _____

3. Forest is to tree as class is to ☐

4. Team is to teem as feat is to ☐

5. Principle is to principal as arc is to ☐

6. ⊞ : ■ :: ◯ : _____

7. ⊞ : ⊞ :: ⊞ : _____

8. Eye is to face as window is to ☐

9. Draw is to ward as tops is to ☐

10. Wolf is to dog as lion is to ☐

11. 9 is to 81 as 10 is to _____

12. Whole is to half as ◯ is to _____

Name_____

Puzzle It! Attribute and Pattern Puzzles   16   ©2008 by Incentive Publications, Inc., Nashville, TN.

Puzzle 11: MUG SHOTS

Mrs. Magillicutty described the features of the dog-napper to the police. She claimed that the culprit was clean-shaven, squinty-eyed, scar-faced, black-haired, and male.

Can you find him in the mug shot book?

THE DOG-NAPPER IS . . .

Puzzle 12 — DOES IT FIT?

One figure in each row doesn't fit with the others. Draw a circle around it.

1.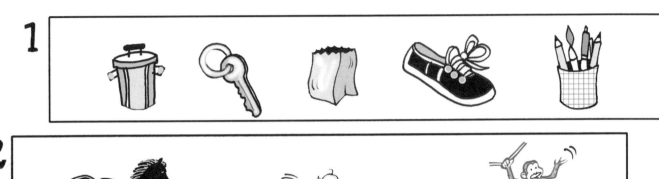

2.

3. 113   97   19   16
       71   23   53   2

4.

5.

BONUS: Look at all five of your circled pictures. Which one doesn't fit with the others? Draw it in the box.

Name _____

Puzzle It! Attribute and Pattern Puzzles

Confusing Confections

Chef Jorge prepared his award-winning torte recipe for his gourmet cooking class. Each student tried to duplicate the dessert down to the last detail. Circle the letters of the dishes that were copied **exactly**.

This is Chef Jorge's Scrumptious Chocolate Ice Cream Banana Torte.

Draw Chef Jorge's torte **exactly**.

Name _____

Puzzle It! Attribute and Pattern Puzzles ©2008 by Incentive Publications, Inc., Nashville, TN.

Going Buggy!

One ladybug is unique.

Can you find it? Circle it and name its unique attribute.

Name_____

Puzzle It! Attribute and Pattern Puzzles   20   ©2008 by Incentive Publications, Inc., Nashville, TN.

PUZZLING PATTERNS #1

1.

Choose **a**, **b**, or **c** to complete the geometric pattern, figure 1.

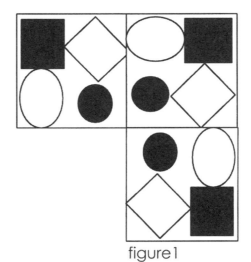
figure 1

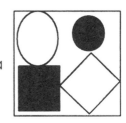
a

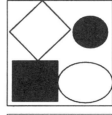
b

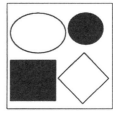
c

---

figure 2a

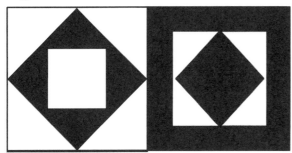

figure 2b

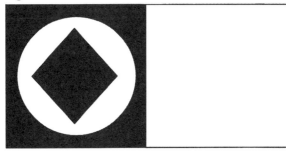

2.

Use figure **2a** as a guide to create the pattern in figure **2b**.

3.

Create a geometric puzzle pattern of your own. Leave one square blank and pass it to a classmate to finish.

figure 3

Puzzle 16 — SEEING DOTS

Someone went wild at the domino factory! See if you can make any sense out of these dotty patterns.
Continue the patterns in each of the four sets of dominoes by drawing the missing dots in the blank domino at the end of each row.

3. Fill in the last two dominoes.

4. Fill in the blank domino with a pattern that matches the other three.

Name_____

Puzzle It! Attribute and Pattern Puzzles

Puzzle 17 — OFF TO THE RACES

There are five races at Mudclop Meadows today.

Complete the number patterns on the horse blankets to discover the number of the winning horse in each race.

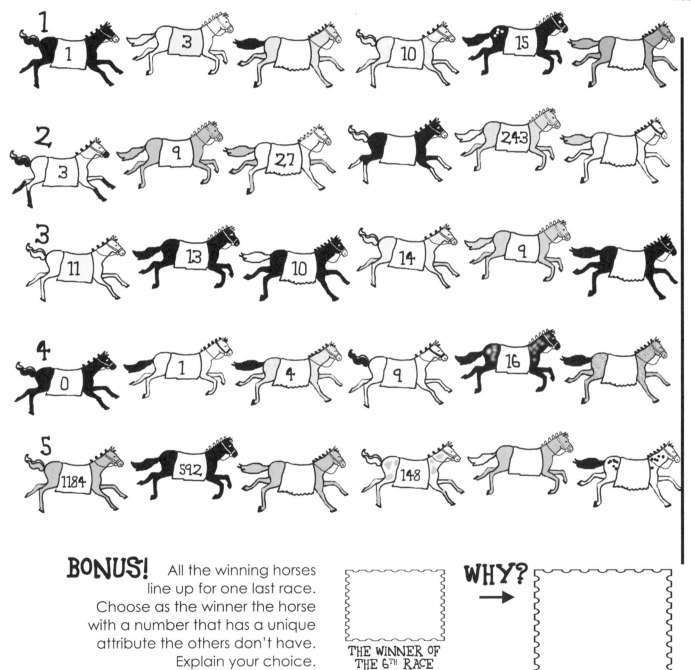

BONUS! All the winning horses line up for one last race. Choose as the winner the horse with a number that has a unique attribute the others don't have. Explain your choice.

THE WINNER OF THE 6TH RACE

WHY? →

Puzzle 18: PUZZLING PATTERNS #2

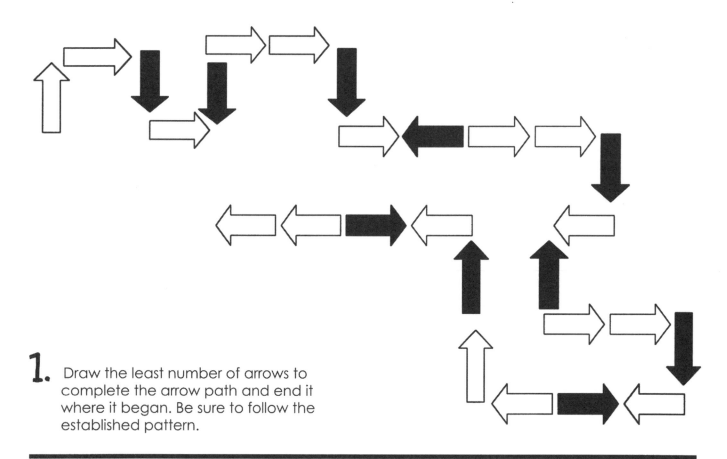

1. Draw the least number of arrows to complete the arrow path and end it where it began. Be sure to follow the established pattern.

---

2. Draw a model of the shapes in the proper arrangement. Or, if you prefer, cut them out and piece them together.

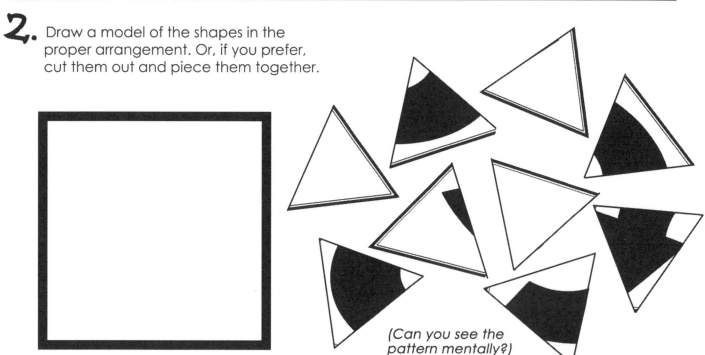

(Can you see the pattern mentally?)

Name _____

Puzzle It! Attribute and Pattern Puzzles

Puzzle 19

The Almost Leaning Tower

Something's wrong with Prince Beauford's new tower. Important pieces are missing! Find the pattern in each section of the tower and draw in the missing pieces.

Hurry, before Beauford's tower comes tumbling down.

HOLEY ATTRIBUTES!

Name _____

Puzzle It! Attribute and Pattern Puzzles ©2008 by Incentive Publications, Inc., Nashville, TN.

Puzzle 20: DOUBLE TAKE

Twenty-two of the words inside the box share an interesting attribute. Take a **second look** and try to figure out what it is.

What is the attribute shared by most of the words? _____

---

refuse   object   project   HEARS

READ   BOW   present   ONE   row

AND   PERMIT   wise   RESUME   bass

TEAR   lead   relay   two

CLOSE   SEWER   A   desert   WORD

minute   understands   wind   number

convert   man   produce   REBEL

Some words do not share the attribute. Circle them.
Rearrange the circled words and write them on the lines to make an old Yiddish proverb about – **words**!

_____ _____ _____ _____

_____ _____ _____ _____

Yiddish Proverb

Name _____

Puzzle 21

WHO STOLE THE DOUGH?

Chef Jorge got on the bus at 5th Street carrying a sample of his secret sourdough bread starter. He fell asleep. When he awoke at his regular stop on 11th Street (six blocks later), the bus was empty and the dough was gone!

Only five other passengers rode the bus at the same time as the chef. They each got off on different blocks starting at 6th Street.

Study the attributes of each passenger and read the clues to decide which one walked off with the chef's dough. Name the street where the culprit got off the bus.

(Use the boxes below to jot down notes as you solve the puzzle.)

UMA   EDNA   IVAN   ALVIN   OPAL

1. The thief did not get off on 7th Street.
2. The person with freckles got off the bus one stop before Chef Jorge.
3. There is a hobby shop specializing in model trains on 7th Street.
4. The person who got off on 6th Street just got braces.
5. The thief does not have a pet.
6. There is a One Hour Photo Developing shop on 8th Street.
7. Uma got off the bus before the thief.

THIEF:

STREET:

Puzzle 22: Riddle Me an Attribute

Identify the attributes described in these riddles.
Use the attributes as clues to help you solve the riddles.

1. I have a head.
   I have a bottom.
   I run in summer,
   Spring and autumn.
   I have no money in my bank.
   I have a mouth, but never drank.
   What am I?

   _____

2. I carry the weight of my world as I roam.
   Wherever I wander I never leave home.
   What am I?

   _____

3. **The Riddle of the Sphinx**
   I go on four limbs in the morning,
   two at noon, and three at twilight.
   What am I?

   _____

4. I run all day
   But never alone.
   My tongue hangs out
   When I get home.
   What am I?

   _____

5. Though I have enlightening powers,
   The length of my life is hours.
   Tall to begin, shorter I grow.
   A gusty wind is my worst foe.
   What am I?

   _____

6. You can't see me
   But you know I'm there.
   I have no substance,
   I am full of air.
   If I make you stumble,
   I'll ruin your day,
   But if you fill me,
   I'll go away.
   What am I?

   _____

A Classic Riddle:

What is black and white and red all over?
(Can you think of four different things with those attributes?)

_____ _____

_____ _____

7. I must be measured for you to know me.
   I pass too fast, or else too slowly.
   What am I?

   _____

Name _____

Puzzle 23
Spy Talk

Stan, a super-sneaky spy, believes he has succeeded in stealing the secret formula for tepid fusion. (His information tells him that this is a breakthrough discovery in sub-atomic energy and that the formula is worth millions.) The document may look like random squiggles and swirls, but there is a pattern hiding here.

Crack the code and write the message.

HEH, HEH!

[Coded message with symbols interspersed with the letters: A...k...S...m...J...bp...O...u...z...R...d...n...d...R...M...Y...m...S...e...T...J...A...J...N...p...N...J...O...s...J...W...T...J...M...H...r...M...E...R...J...F...O...E...K...A...E...S...aN...Z...O...K...N...C...Y...v...J...O...D...U...]

TALKING PATTERNS

PEOPLE ARE TALKING ABOUT THESE ITEMS IN THE NEWS!

Some of the blocks in the crossword puzzle have figures inside. These figures match letters in the alphabet box below. Find the letters and write them beside the matching blocks in the crossword.

Read the crossword clues and figure out what words the blocks are spelling. Fill in the symbols that represent the missing letters.

CLUES

DOWN
1. A NON-RENEWABLE ENERGY SOURCE
2. CYBER NETWORK
3. ENVIRONMENTAL ISSUE
4. HOMEWORK HELPER

ACROSS
5. SOME CELEBRITIES PURSUE A WILD AND CRAZY _____.

Alphabet Box

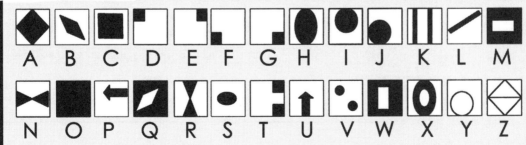

Name _____

Puzzle It! Attribute and Pattern Puzzles   30   ©2008 by Incentive Publications, Inc., Nashville, TN.

Puzzle 25 — Fly Trap

Mr. Spider must follow a path of words that share a particular attribute. Show him the way by drawing a line from where he is now to the delicious fly that awaits him.

Use the lines at the bottom to explain why you chose a certain path.

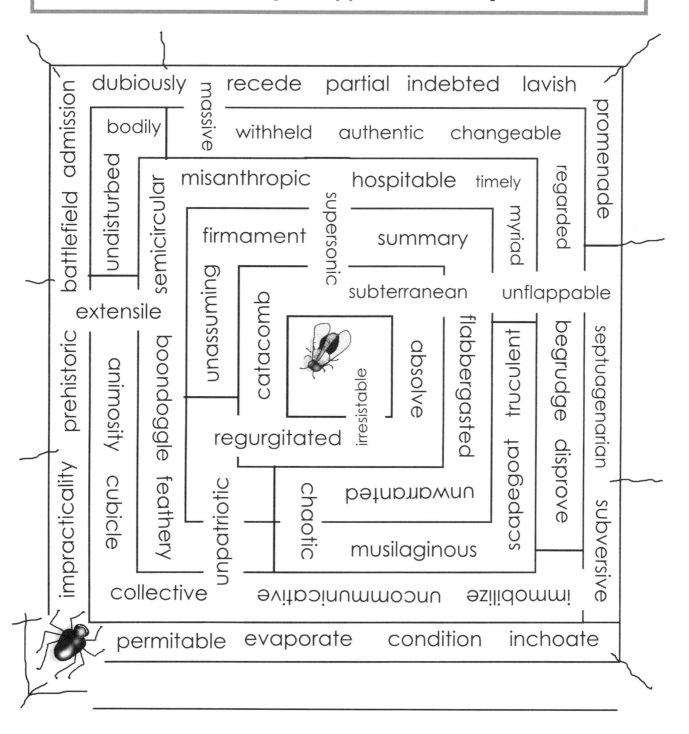

Name _____

MIRRORED PATTERNS

Study the random pattern of black and white blocks in section 1 of the figure below.

Use a black marker or pencil to create its mirror image in section 2. Then, create the mirror image of sections 1 and 2 in sections 3 and 4.

Puzzle 28: Logo Logistics

The Acme Apex Ad Agency and the Dazzle-Flash Media Group have each submitted a portfolio of logo ideas to a big corporation.

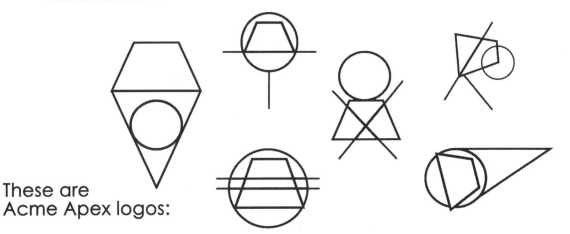

These are Acme Apex logos:

These are Dazzle-Flash logos:

Circle the logos that are neither Acme Apex nor Dazzle-Flash.

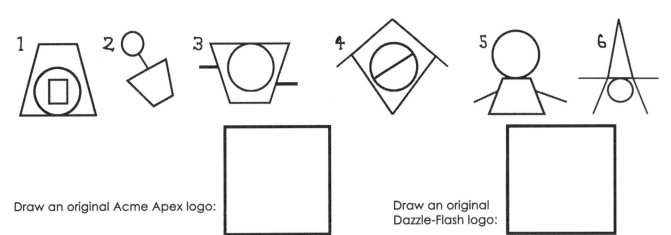

Draw an original Acme Apex logo:

Draw an original Dazzle-Flash logo:

Name _____

Puzzle It! Attribute and Pattern Puzzles 34 ©2008 by Incentive Publications, Inc., Nashville, TN.

Puzzle 29
GERM-INATIONS

Teenaged scientists have discovered microorganisms that replicate themselves in a strange new way. The slides below show several parent-child pairs.

These slides are examples of the new discovery.

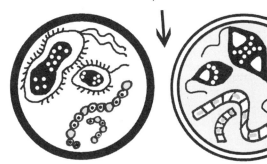

These slides are not.

Circle the letters of the slide specimens that have the **attribute** the scientists discovered.

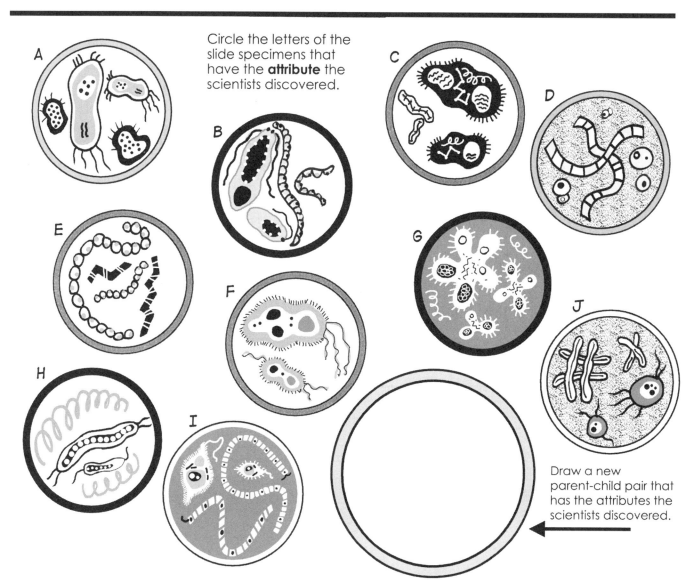

Draw a new parent-child pair that has the attributes the scientists discovered.

Name _____

Puzzle It! Attribute and Pattern Puzzles ©2008 by Incentive Publications, Inc., Nashville, TN.

Puzzle 30: ATTRIBUTES WITH "CLASS"

Most of the teachers at Thorny Hollow Middle School have different seating charts for their classes. Look at the ways four teachers have organized the same five students.

Each teacher used one attribute to determine the class list order. Study the lists to determine attributes they used. Write the attributes under the appropriate class list.

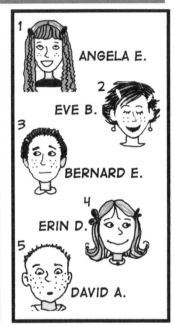

First Period
Mr. Witt

Second Period
Miss Smiley

Third Period
Mrs. Earnest

Fourth Period
Mr. Graves

_____   _____   _____   _____

Arrange these five students in three different ways using attributes having to do with their appearance. Write a different attribute above each box. List the names in a logical order.

ANGELA   BERNARD   EVE   ERIN   DAVID

1 _____
2 _____
3 _____
4 _____
5 _____

1 _____
2 _____
3 _____
4 _____
5 _____

1 _____
2 _____
3 _____
4 _____
5 _____

Name _____

Puzzle It! Attribute and Pattern Puzzles

Puzzle 31 — Overlapping Pattern Chains

Four different pattern chains overlap. Each chain has a unique pattern. Be careful, these can be tricky.

1. Complete pattern chains **1a** and **1b**.

2. Study the pattern in chain number **2**. Describe the pattern sequence on the lines.

_____
_____
_____
_____
_____

3. Draw a line under chain number **3** from beginning to end.

   Use a pencil to complete the last six squares in the pattern.

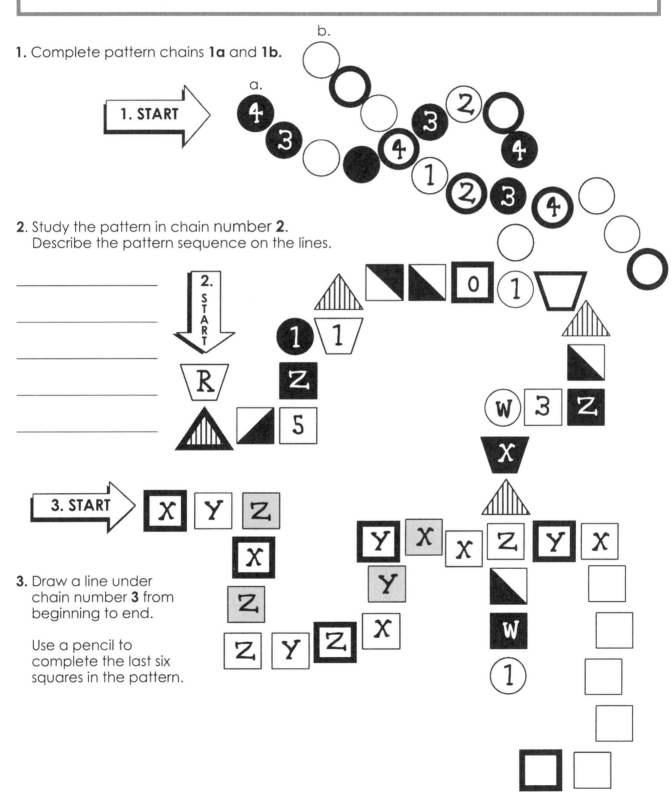

Name _____

Puzzle It! Attribute and Pattern Puzzles 37 ©2008 by Incentive Publications, Inc., Nashville, TN.

Puzzle 32
POINTS OF STYLE

Thorny Hollow Middle School has designated next Friday as **Wild and Crazy Costume Day**. The costumes the students are allowed to wear must meet certain requirements. Read the flyer that the principal posted on the bulletin board. Create a costume that has all the attributes described.

Wild and Crazy Costume Attributes:

1. Must have three (or more) parts
2. Must have two (or more) patterns
3. Must use a minimum of four colors
4. Masks optional
5. May use wigs, hats, or hair decorations
6. Cannot be wider (or taller) than a door
7. Must have one accessory
8. May be any living being

LIST THE SPECIAL ATTRIBUTES YOU ADDED THAT ARE NOT ON THE LIST:

DRAW YOURSELF WEARING THE COSTUME YOU'VE DESIGNED. MAKE SURE IT HAS ALL THE RIGHT ATTRIBUTES. ADD SOME OF YOUR OWN.

Name

Puzzle It! Attribute and Pattern Puzzles ©2008 by Incentive Publications, Inc., Nashville, TN.

Puzzle 33
INTERESTING CHOICES

1. COSTUMES MUST HAVE TWO PATTERNS.
2. COSTUMES MUST HAVE THREE COLORS, OR SHADES.
3. COSTUMES MUST HAVE SYMMETRY.
4. COSTUMES MUST BE A PERSON, ANIMAL, OR THING WITH AN <u>UNDERWATER</u> THEME.

The Science Club is sponsoring a dance with a prize for the best costume. The winning costume must meet all the requirements listed above.

Find the most complete and accurate costume and award First Prize to that student. Then, award a joke prize to the kid with none of the attributes listed.

Write the winning names in the boxes.

FIRST PLACE:

PRIZE:

LAST PLACE:

PRIZE:

BONUS! Unscramble the letters in both names and find their prizes.

Just for Fun!
Which costume do you think was the most creative (even if it couldn't win the prize)?

Name_____

Puzzle It! Attribute and Pattern Puzzles   39   ©2008 by Incentive Publications, Inc., Nashville, TN.

Puzzle 34: Number Puzzlers

ONE MINUTE TEASER

Angie's shopping list:

Dog food _____ $4.10
Leash _____ $6
Flea powder ____ $2.75
Chew toys ___ 3 for $4
Caring for Puppies
Handbook _____ $13.90

Simon's shopping list:

Dog food ------ $12.90
Leash --------- $5.55
Flea powder --- $6.92
Chew toy ------ $4.25
Caring for Older Dogs
Handbook ------ $12.77

Angie is a careful shopper. She likes to buy items that have a certain attribute. The items on Simon's list share one attribute that is important to Angie. What is that attribute? What else do you know about Angie that is different than Simon?

BRAIN TWISTER!

Max loves to organize. He even organizes his addition problems. All his favorite problems have the same pattern of attributes. Figure out what these attributes are and write a new problem for Max on the empty page.

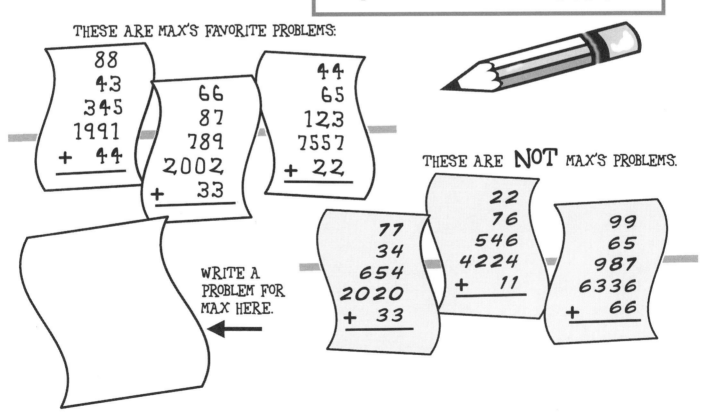

THESE ARE MAX'S FAVORITE PROBLEMS:

```
  88        66         44
  43        87         65
 345       789        123
1991      2002       7557
+ 44     +  33       + 22
```

THESE ARE **NOT** MAX'S PROBLEMS.

```
  77         22         99
  34         76         65
 654        546        987
2020       4224       6336
+ 33      +  11       + 66
```

WRITE A PROBLEM FOR MAX HERE.

Name _____

Puzzle It! Attribute and Pattern Puzzles

Puzzle 35: The Catch of the Day

Finn has netted an odd assortment of objects today. When he looks closely, he sees that the items in each net are actually connected in a certain way.

For each net, find the attribute that is common to all the items. Write the common attribute underneath each net.

A
169   16,900   65
          26
182   130
   325   117

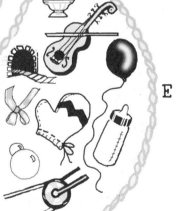

, # Puzzle 36 — Puzzling Patterns #3

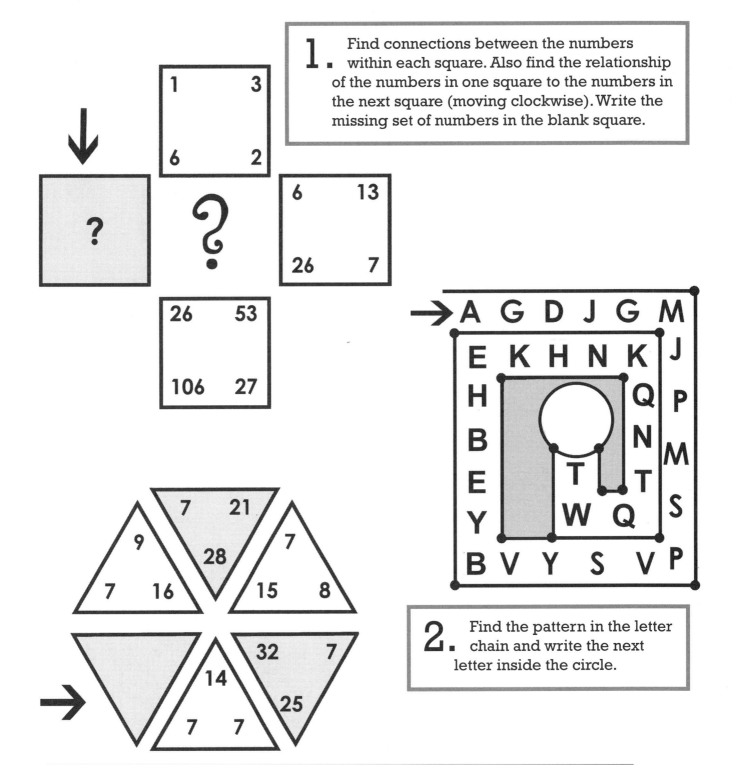

1. Find connections between the numbers within each square. Also find the relationship of the numbers in one square to the numbers in the next square (moving clockwise). Write the missing set of numbers in the blank square.

2. Find the pattern in the letter chain and write the next letter inside the circle.

3. Each segment of the puzzle has a similar pattern of numbers. Find out what it is and write the missing numbers in the blank triangle.

Name _____

Puzzle It! Attribute and Pattern Puzzles   42   ©2008 by Incentive Publications, Inc., Nashville, TN.

Puzzle 37
Stuffed Lockers

Katelyn, Burke, Ruby, Alex, and Annabeth have adjoining lockers. They all have unusual items in their lockers. One of these is a smelly turkey. Follow the clues to figure out which locker belongs to each person, and whose locker holds the turkey.
Use the boxes below to write the number of each student's locker.

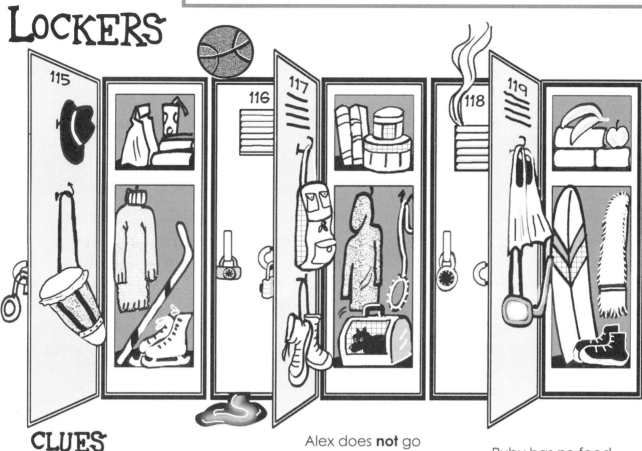

CLUES

Annabeth is **not** a basketball player.

Alex does **not** go near ice rinks or bodies of water.

Ruby has **no** food in her locker.

Burke's locker is **not** next to a locker with a missing lock.

The locker with the turkey also has a deflated raft.

Katelyn's locker does **not** have a drum.

Annabeth has food in her locker.

Alex's locker is between Ruby's and Burke's.

Ruby's locker is next to locker 118.

Soup is spilled in Alex's locker.

The basketball player's locker is two lockers away from the locker with the raft.

Katelyn's locker has an odd number.

ANNABETH

BURKE

KATELYN

ALEX

RUBY

Puzzle 38: Cross-Number Puzzle

The solutions to the math problems are already in the puzzle. All you have to do is decipher the code! Find the matching number for each crossword symbol in the key below. Write the number near its symbol in the puzzle.

Solve the problems. As you find the correct answers in the puzzle, write the letter of the problem on the lines below. *(The first one is done for you.)*

a. (27 + 45 - 19) × 4 - 2 = _____
b. 12,437 + 5632 + 40,934 + 3030 = _____
c. 9.26 × 82 = _____
d. 987,996 - 56,765 = _____
e. 15 × 10 × 9 × 2 = _____
f. 2005 + 2006 + 2007 + 2008 = _____
g. 111 × 2 = _____
h. 12,000 divided by 6 equals _____

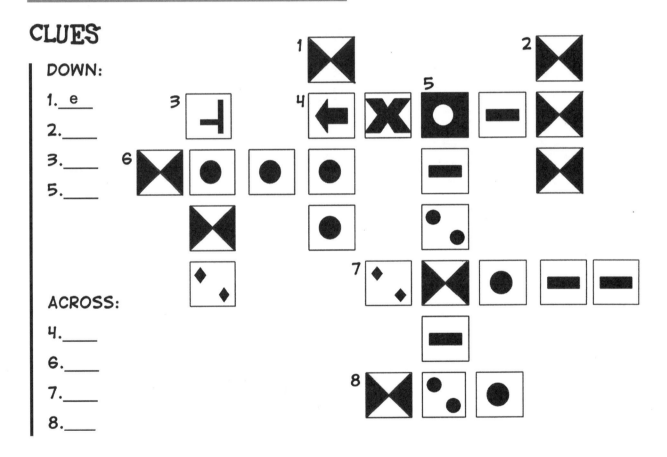

CLUES

DOWN:
1. _e_
2. ___
3. ___
5. ___

ACROSS:
4. ___
6. ___
7. ___
8. ___

BONUS: What attributes (if any) do the answers to the problems share? _____

Puzzle 39 LIFE IN A FISH BOWL

THESE ARE RANDY'S FAVORITE FISH.

THESE ARE NOT RANDY'S FAVORITES.

Darcy wants to buy some fish for Randy's birthday. Which of the fish in the Pet Shop tank should she choose? Circle the numbers of the fish that have the attributes shared by all of Randy's favorite fish.

PET SHOP

Name_____

Puzzle It! Attribute and Pattern Puzzles ©2008 by Incentive Publications, Inc., Nashville, TN.

Toss the Dice

The first two figures in each section show different views of the same cube (or die). There are no two sides alike on any of the figures. Study the views that are shown, then decide what images are on the blank sides of the third view in each row.

1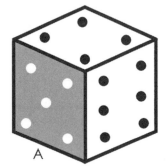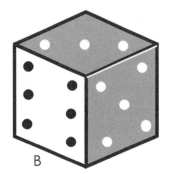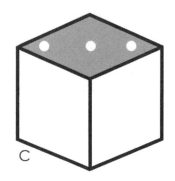

DRAW THE UNSEEN SIDES.

2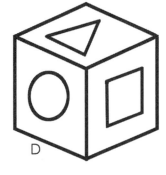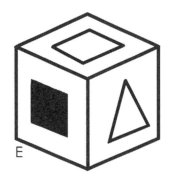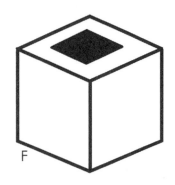

DRAW THE UNSEEN SIDES.

3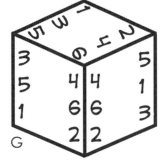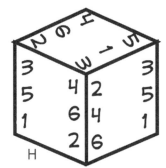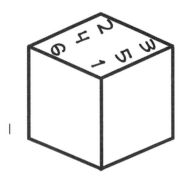

DRAW THE UNSEEN SIDES.

Name_____

Puzzle It! Attribute and Pattern Puzzles   46   ©2008 by Incentive Publications, Inc., Nashville, TN.

Puzzle 47 — Rah-Rah Combos

The cheerleading squad is practicing for the big game. Some of their routines are eye-catching!

1. Use the pattern of cheerleaders in groups 1, 2, and 3 as a model. How many more arrangements can these six cheerleaders make with **B** as the base? _____

2. How many arrangements can be made by substituting three **A** cheerleaders as the base with **B**, **K**, and **M** on their shoulders? _____

3. Think about the order of cheerleaders in towers 1, 2, 3, and 4. Finish the last two towers by writing the letters of the missing cheerleaders inside the empty boxes.

4. Which cheerleader from the group on the right will complete the pattern? Write the letter in the box.

Name _____

PUZZLING PATTERNS #4

1. Each of these black and white patterns has a negative match somewhere in the group. Draw lines to make the matches.

a   b   c   d

e   f   g   h

2. In your mind, hold one end of this chain and let the links fall into a straight line. Write the letters of the links in order.

Hint: Start with the link nearest the top that has only one connecting link.

1 _____

2 _____

3 _____

4 _____

5 _____

6 _____

7 _____

8 _____

Name_____

Puzzle It! Attribute and Pattern Puzzles

Puzzle 43 — Mini-Mysteries

See how quickly you can solve these rapid-fire challenges.

1.

Tyra is ahead of Melanie but behind Emily. Kira is ahead of Shelly and Emily. Melanie is ahead of Shelly. Which runner is 110?

2.

In a dog show, the second place prize went to one of the taller dogs.
Dog number 6 placed higher than number 12 and lower than number 7.
Dog number 3 won third prize.
Dog number 21 placed higher than number 6.
Which dog placed first?

3. Five parcels arrive in this order on the post office conveyor belt.

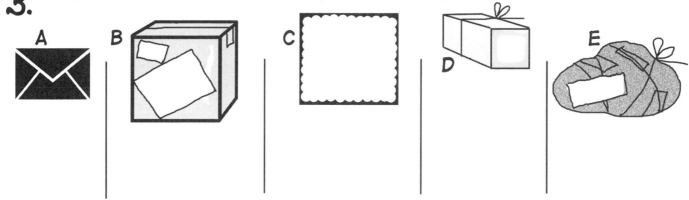

Each one is from a different city and holds different contents. The package from Boise holds a valuable five-page document. Packages **D** and **C** have food. The sausages from Chicago are not next to the stuffed skunk from St. Louis. The package from Chicago is between the chocolate cake crumbs from Tucson and the package from Albany. The documents are two packages away from the cake crumbs. The stuffed skunk is next to the documents. Which package holds the diamonds and from what city did it come?

Puzzle 44: Analogy Quick Thinks #2

Choose or draw the figure that completes each analogy correctly.

1. Finish the analogy.

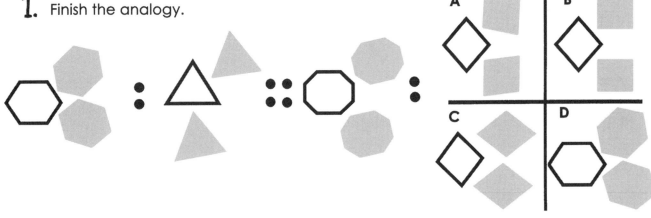

2. Finish the analogy.

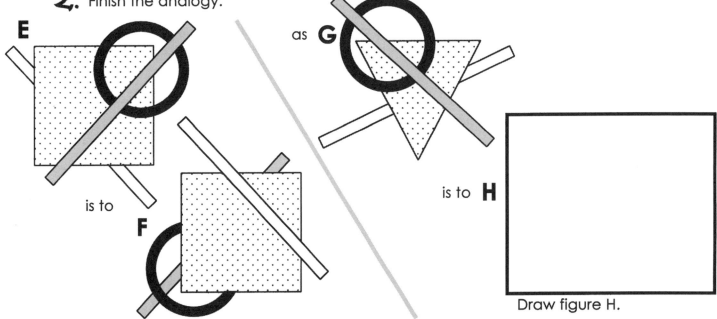

Draw figure H.

3. Choose the best figure to complete the analogy.

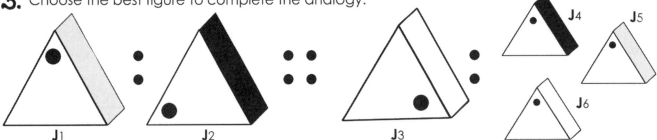

Puzzle 45: Pizza Portions

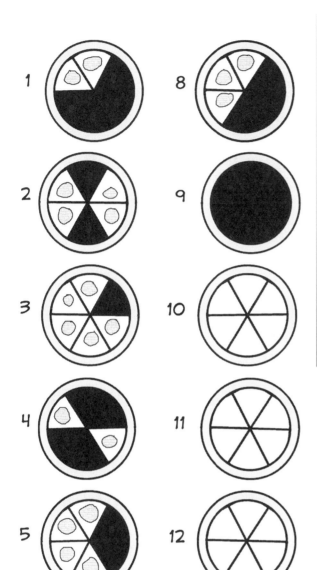

Chef Jorge treated the fourteen members of the debate team to individual-sized pizzas cut into six equal slices.

Amazingly, when the students were full and stopped eating, each of the fourteen pizza trays has a different configuration of leftover pizza slices!

Study the patterns. Which arrangements of pizza slices are not shown? Draw the correct number and placement of the missing pizza slices on platters 10 through 14.

Hint: If the leftover patterns have the same number of pieces, they should never be arranged exactly the same way, even if the trays are rotated.

LEFTOVERS, ANYONE?

Bonus: How many full boxes of leftover pizza did Chef Jorge send back to the school?

Name_____

Puzzle It! Attribute and Pattern Puzzles

Puzzle 46 — PRIZEWINNING NUMBERS

1. Three attributes make the **A** numbers prize-winners in a contest.

The numbers in **B** are not prize-winners.

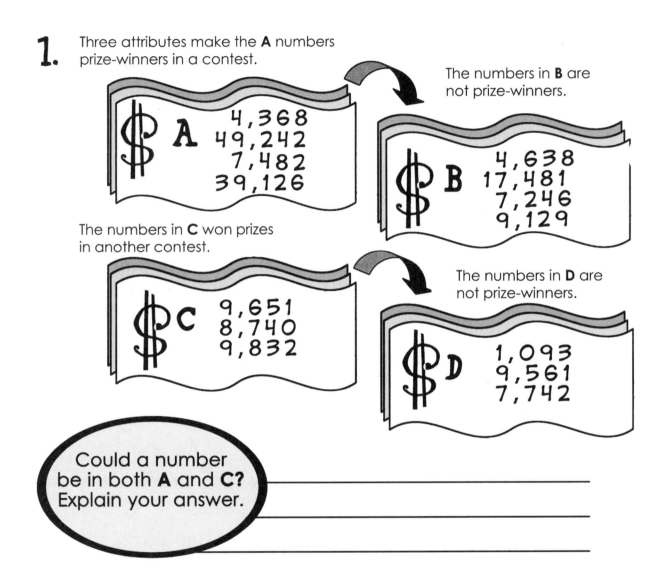

A: 4,368; 49,242; 7,482; 39,126

B: 4,638; 17,481; 7,246; 9,129

The numbers in **C** won prizes in another contest.

C: 9,651; 8,740; 9,832

The numbers in **D** are not prize-winners.

D: 1,093; 9,561; 7,742

Could a number be in both A and C? Explain your answer. _____

---

2. Can you think of any numbers that have the following attributes?

- no even digits except 0
- 3–5 digits
- hundreds digit is 4 less than the ones digit
- sum of all digits is 18

Write your answers here.

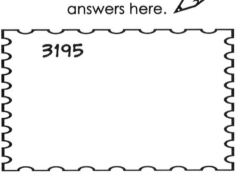

3195

Name _____

Puzzle It! Attribute and Pattern Puzzles   52   ©2008 by Incentive Publications, Inc., Nashville, TN.

Puzzle 47 — PREDICTABLE PATTERNS

Predict the patterns on these unusual license plates.

1 Figure out the pattern and fill in the missing numbers and letters in a, b, and c.

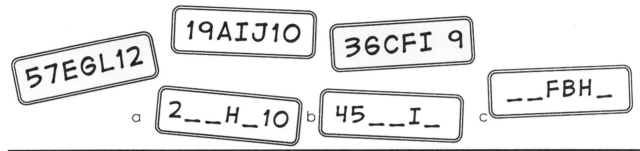

2 Figure out the code on these vanity plates. Write the answers in the lower set of plates.

(Example:)
3OJHIU
2NIGHT

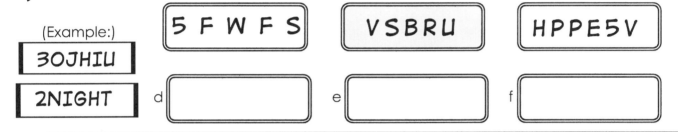

d ____ e ____ f ____

3 This set of plates follows a pattern that describes a well-known set of facts. Discover the pattern and write the correct letters and numbers on the last two license plates.

JA31   FE28   MA31   AP30

g ____ h ____

4 Finish the analogy.

5 Figure out the puzzles.

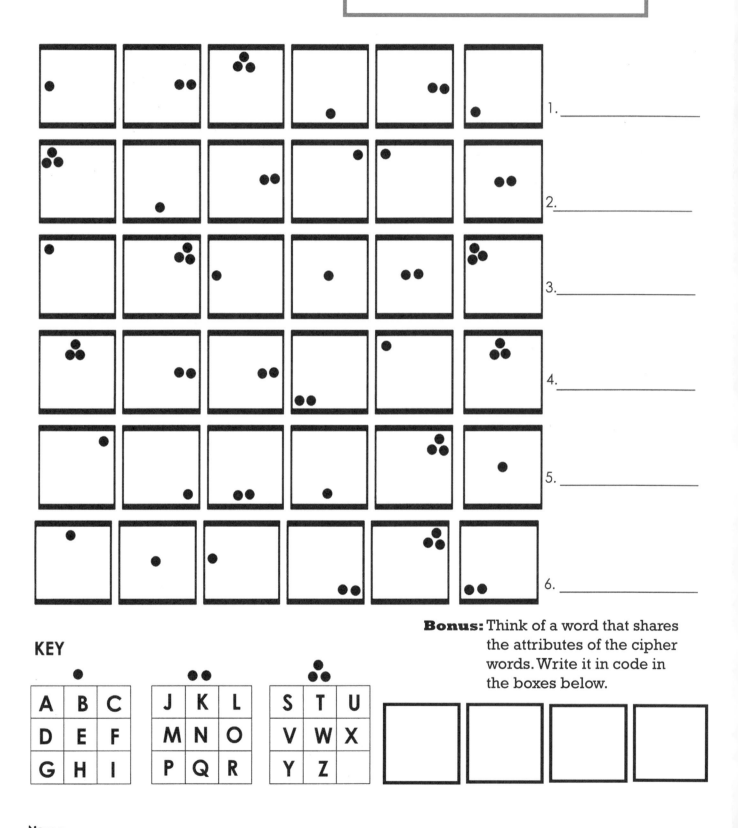

Puzzle 49 — Book Bewilderment

The school librarian has a shelf for all his favorite books. These are librarian Higgle's favorite books.

SHELF A

These are not books the librarian shelves with his favorites.

SHELF B

1. _____
2. _____
3. _____
4. _____
5. _____
6. _____

These books are waiting to be shelved. Write the letter of shelf (A or B) on the lines beside the books.

What attributes do librarian Higgle's favorite books share?

Name _____

Puzzle It! Attribute and Pattern Puzzles 55 ©2008 by Incentive Publications, Inc., Nashville, TN.

Puzzle 50

A Tribute to Flags

Students at Thorny Hollow Middle School designed these flags.

Thorny Hollow students did not design these flags.

Study both sets of flags. Draw a flag like the ones designed by the Thorny Hollow students. Explain how your flag shares the same attributes as the students' flags.

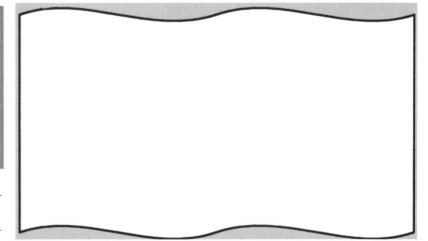

Name _____

Puzzle It! Attribute and Pattern Puzzles 56 ©2008 by Incentive Publications, Inc., Nashville, TN.

Puzzle 51 SQUARED!

Cut out the pieces of the puzzle and arrange them in a pattern on the center square.

Puzzle 52: Eighth Grade Musical!

The student band, Pizazz, will be performing in their middle school's new production, **Eighth Grade Musical**!

At today's rehearsal they decided to play a joke on the music director, Mr. Harmony. They all switched instruments. What a ruckus of sour notes they made!

These are the members of the band:
Moogie, Geneva, J.J. Hipp, Crystal, and Stringbean.

These are the instruments:
saxophone, bass, keyboard, drums, and electric guitar.

Figure out which instruments the students are playing today. Match the correct instrument with each player by writing the name of the instrument in the box under each student's name.

CLUES:

Stringbean is not wearing striped pants.

Today, J.J. is not playing the keyboard.

At today's rehearsal, Geneva is on the opposite side of the stage from Moogie.

Name_____

ANSWER KEY

Puzzle 1 (pg 7)
Kick-Flipping Attributes

The Lightning Bolt skateboards are decorated with patterns using only four lines. There are 6 Lightning Bolt skateboards.

Puzzle 2 (pg 8)
Swing Your Partner

Draw lines to connect partners that will make a sum of 13 dots. Each pair must have one black and one white partner. Each pair must have one female and one male partner.

Puzzle 3 (pg 9)
Snappy Beads

The pattern is: Two white 3-sided figures followed by one black 6-sided figure, two white 4-sided figures followed by one black 8-sided figure. Puzzle-solvers should continue the pattern with two white 5-sided figures followed by one black 10-sided figure, and two white 6-sided figures followed by one black 12-sided figure.

Puzzle 4 (pg 10)
Space Woogies

The attributes of a Woogie are: oval (egg-shaped); 2 legs; 3 eyes; 4 arms; 4 fingers.

Puzzle 5 (pg 11)
Puzzling Discs

Esteban's titles all have a participle (verb ending in *ing*).
Sophia's titles all contain at least one preposition.
Devon's titles each have three syllables. The last pictured CD would NOT fit into Devon's collection.

Puzzle 6 (pg 12)
Note-ables

1.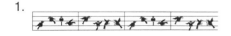

2.

3.

4. Make sure students create a repetitive pattern.

Puzzle 7 (pg 13)
Quick Thinks

1. rectangle lying on its long side with dots; white background with dots.
2. triangular prism, white with right side face shaded.
3. answer C.
4. The blocks drawn must be a quarter-turn (clockwise) rotation of the left hand block set. Block #1 will be on the left; blocks 2 and 3 on the right.
5. Puzzle-solvers must draw the mirror image of the shape stack and rearrange the letters to match the first set. The letters will spell *symbols*.

Puzzle 8 (pg 14)
Making Matches

Picture groups:
tree O, pencil O, fence D, log cabin W. The word is WOOD.
Liberty Bell E, doorbell L, jingle bells B, alarm bell L. The word is BELL.
shark F, goldfish I, swim fins N. The word is FIN.
push-pin H, scissors A, safety pin S, dart R, knife P. The word is SHARP.

Puzzle 9 (pg 15)
Carousel Comparisons

Karina's new horses must always have the right hind leg down and in front of the left hind leg, the front left leg bent, three up-turned curls in the tail, and three down-turned curls in mane and forelock. Carousel horses 2, 3, 4, 8, and 9 are Karina's new horses.

Puzzle 10 (pg 16)
Picture This!

1. car ; 2. 11 ; 3. student (single); 4. feet; 5. ark; 6. figure of quarter circle wedge; 7. figure with dark square in upper right corner; 8. house, (school, windmill, and ark also acceptable); 9. stop sign; 10. cat; 11. 100; 12. half circle figure

Puzzle 11 (pg 17)
Mug Shots

The culprit is the only face with all five attributes: #1121.

Puzzle 12 (pg 18)
Does It Fit?

1. **key** (the only non-container)
2. **Skateboard** (the only non-living thing)
3. **16** (the only non-prime number)
4. **Steak** (the only non-fruit or vegetable)
5. **the stop sign** (the only item that doesn't make a noise)

BONUS– **key**. It is the only one of the answers that doesn't begin with the letter **s**.

Puzzle 13 (pg 19)
Confusing Confections

Only letters J and K copy the chef's dessert exactly. Make sure the drawings replicate the chef's dessert precisely.
(Note: position of stem of top cherry; size of fudge topping; width of whipped cream layer; number of sprinkles; location of white and dark ice cream scoops; number of cherries on bananas; position of cherry stems on lower layer—on top of banana slices.)

Puzzle 14 (pg 20)
Going Buggy!
A single ladybug is the only asymmetrical bug on the page. (It has a different number of spots on its two halves.) To find the asymmetrical ladybug count (from the lower left hand side) five bugs up and one to the right.

Puzzle 15 (pg 21)
Puzzling Patterns # 1
Figure 1: b is the match.

Figure 2: The second part of 2b must show a white background holding a large black diamond with a smaller white circle inside the diamond.

Figure 3: Make sure each puzzle-solver creates a true pattern with at least 2 attributes.

Puzzle 16 (pg 22)
Seeing Dots
1. Answer on blank domino: 1 over 2.
2. Answers on blank domino: 0 over 24.
3. "Pairs" of successive dominoes (1 and 2, 3 and 4, 5 and 6) create the basic pattern: first domino, top half is 1; second domino, bottom half is 2; second domino, top half is 3; first domino, bottom half is 4. The 4 becomes the top half of the third domino, 5 the lower half of the fourth domino, 6 the top half of the fourth domino, 7 the lower half of the third domino, and so forth. So, the answer for the set of blank dominoes is:

10	12
13	11

4. The last domino has three possible correct answers: 2 and 8; 4 and 6; or 0 and 10. The dots on the two sides of the domino must have a sum of ten and the number of dots on the left side must be less than or equal to the number of dots on the right side.

Puzzle 17 (pg 23)
Off to the Races
Race 1: 1; 1 + **2** = 3; 3 + **3** = 6; 6 + **4** = 10; 10 + **5** = 15; 15 + **6** = answer: **21**

Race 2: 3; 3 x 3 = 9; 9 x 3 = 27; 27 x 3 = 81; 81 x 3 = 243; 243 x 3 = **729**

Race 3: 11; 11 + 2 = 13; 13 − 3 = 10; 10 + 4 = 14; 14 − 5 = 9; 9 + 6 = **15**

Race 4: 0^2 = 0; 1^2 = 1; 2^2 = 4; 3^2 = 9; 4^2 = 16; 5^2 = **25**

Race 5: 1184 divided by 2 = 592; 592 divided by 2 = 296; 296 divided by 2 = 148; 148 divided by 2 = 74; 74 divided by 2 = **37**

BONUS: Answers may vary. One correct answer is 37, the only prime number, but puzzle-solvers may identify other unique numbers.

Puzzle 18 (pg 24)
Puzzling Patterns # 2
Figure 1: The pattern is a path of arrows: two white arrows, followed by one black arrow, followed by one white arrow, followed by one black arrow pointing back to the last white arrow. The puzzle requires at least three arrows to finish—one black arrow facing left, one white arrow facing left, and one black arrow pointing backwards above the end of the previous white one and below the beginning arrow.

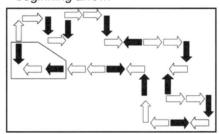

Figure 2: Nine small triangles are randomly arranged. When pieced together correctly they form a larger triangle with a curved arrow inside. Students can solve this mentally, can recreate by drawing a model, or can cut out and piece together actual paper triangles.

Puzzle 19 (pg 25)
The Almost Leaning Tower
Puzzle-solvers will do their best to recognize and copy the patterns.

Puzzle 20 (pg 26)
Double Take
All but nine of the words are homographs—words that look the same (same spelling), but have different pronunciations and different meanings. The words that do not have this attribute are: *hears, one, and, wise, two, a, word, understands,* and *man.*

The Yiddish proverb unscrambled: A wise man hears one word and understands two.

Puzzle 21 (pg 27)
Who Stole the Dough?
The thief is Ivan. He got off the bus on 9th Street.

Puzzle 22 (pg 28)
Riddle Me an Attribute
1. a river 2. a snail, or a turtle 3. a human being as he ages 4. a shoe 5. a candle 6. a hole 7. time 8. Answers will vary (a blushing zebra, or sunburnt penguin, or panda eating strawberries; a newspaper, etc.).

Puzzle 23 (pg 29)
Spy Talk

The pattern (⌘⌘≈⌘✗⌘●) appears regularly in the larger cipher. This pattern is always followed by a letter. These letters placed chronologically compose the message:

SORRY STAN NOW THE JOKE'S ON YOU!

Puzzle 24 (pg 30)
Talking Patterns

DOWN:
1. fossil fuel
2. World Wide Web
3. climate change
4. computer

ACROSS:
5. lifestyle

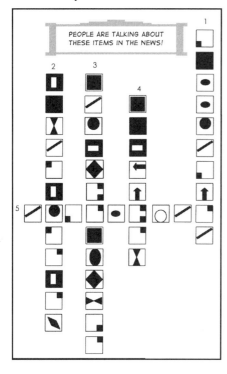

Puzzle 25 (pg 31)
Fly Trap

The correct path of words must have both a prefix and a suffix. Start with *impracticality*; move on to *prehistoric; extensile; semicircular; misanthropic; supersonic; subterranean; unflappable; septuagenarian; subversive; immobilize; uncommunicative; unpatriotic; regurgitated; irresistible*.

Puzzle 26 (pg 32)
The Eye of the Beholder

Examples 1, 2, and 3 (Van Pablo paintings) have a circle, a blotch, and five straight line segments, three of which make up an equilateral triangle. The lines do not extend beyond the vertices of the triangle.

Examples 4, 5, and 6 (fakes) each have one circle, one blotch and five line segments, but none have an equilateral triangle.

Number 7 is a fake. Puzzle-solvers should be able to say why. (It has a blotch, a circle, and five straight line segments, but the line segments extend beyond the vertices of the equilateral triangle.)

Students should be able to explain how their drawings fit into the "original" or "fake" category.

Puzzle 27 (pg 33)
Mirrored Patterns

Puzzle 28 (pg 34)
Logo Logistics

All the Acme Apex logos are constructed of a trapezoid, a circle, and two additional line segments. The additional line segments may form another figure.

All the Dazzle-Flash logos are constructed of one circle and five additional line segments. The additional line segments may form another figure.

Logo 1 is neither Acme nor Dazzle.

Make sure puzzle-solvers create accurate logos with above attributes.

Puzzle 29 (pg 35)
Germ-inations

Puzzle-solvers must perceive that all the new specimens are precise, half versions of the originals (half the lines, half the dots, half the squiggles, etc.), not the mini-clones or random pairs shown in the wrong samples. The circled letters should be B, C, E, H, I, J. Make sure drawings contain the attributes of the first two slides.

Puzzle 30 (pg 36)
Attributes with "Class"

Class seating orders:

Mr. Witt: alphabetical order.

Miss Smiley: number of letters in name, least to most.

Mrs. Earnest: Group One, names with the same letter beginning and ending. Group Two, no attribute.

Mr. Graves: The initial of a student's last name is the first letter of the next student's first name.

At the bottom, make sure that puzzle-solvers make logical (or chronological) lists. Some attributes that might be used: number of freckles, length of hair, position of eyes, size of smiles.

Puzzle 31 (pg 37)
Overlapping Pattern Chains

1. Pattern a is a number pattern: 4–3–2–1
   Pattern b is a number pattern: 1–2–3–4

2. A pattern sequence includes different shapes: trapezoid, triangle, 3 squares, circle. Repeats.

3. The pattern involves color and line: White square (thick line), white square, gray square, white square (thick line), gray square, white square, white square. Repeat.

Puzzle 32 (pg 38)
Points of Style

Make sure the costumes drawn meet all the requirements and that the additional attributes added to any costumes are well described.

Puzzle 33 (pg 39)
Interesting Choices

Meg O' Davie has the only costume with all the attributes. Unscrambled, her name spells *video game*. Last place prize is Rick Ebern, who has none of the attributes. His name unscrambled spells *rubber chicken*.

Puzzle 34 (pg 40)
Number Puzzlers

Top half: the prices of Angie's preferred items are always multiples of five. Angie has a puppy.

Bottom half: Stan's pattern: Line 1 is an even two-digit number with a repeated digit. Line 2 is two consecutive digits in reverse order. Line 3 is three consecutive digits, each one greater than the one to the left. Line 4 is a four digit palindromic number. Line 5 is a two-digit number one half the value of line number one.

New problems will vary.

Puzzle 35 (pg 41)
The Catch of the Day

In the nets:
A. All the numbers are multiples of 13.
B. The common attribute is the prefix referring to a number: Uni- corn; bi- cycle; tri- angle; quarter; quintet. Prefixes represent 1, 2, 3, 4 and 5.
C. The common attribute is "ea" in the word represented by the picture: bread, beard, heart, bear, ear, earth, seat, bean, breath, and wreath.
D. The common attribute is "oo" in the word represented by the picture: foot, igloo, balloon, spoon, boomerang, moose, broom.
E. The common attribute is a double consonant in the middle of the word represented by the picture: dessert, fiddle, tunnel, ribbon, mitten, bottle, balloon, pulley, bubbles, button, apple.
F. The common attribute is the number of letters (4) in the word represented by each picture: nine, bell, shoe, five, nose, plus, hand, ring, four, knee, door, book.

Puzzle 36 (pg 42)
Puzzling Patterns # 3

1. Blank square: upper left corner, 106; upper right corner, 213; lower right corner, 107; lower left corner, 426.
2. Z
3. Blank triangle: upper left corner, 23; upper right corner, 30; bottom angle, 7.

Puzzle 37 (pg 43)
Stuffed Lockers

By filling in information in a chart such as the one shown below, possibilities can be eliminated one by one, finally revealing the ownership of each locker. Locker 115 is Burke's, 116 is Alex's, 117 is Ruby's, 118 is Annabeth's, and 119 (Katelyn's locker) holds the turkey.

Locker:	115	116	117	118	119
Katelyn	no	no	no	no	**YES**
Burke	**YES**	no	no	no	no
Ruby	no	no	**YES**	no	no
Alex	no	**YES**	no	no	no
Annabeth	no	no	no	**YES**	no

Puzzle 38 (pg 44)
Cross-Number Puzzle

a. 210 e. 2700
b. 62,033 f. 8026
c. 759.32 g. 222
d. 931,231 h. 2000

DOWN:
1. e 2. g 3. f 4. d

ACROSS:
4. c 6. h 7. b 8. a

Shared attribute: All solutions have 2 as at least one of the digits.

Puzzle 39 (pg 45)
Life in a Fish Bowl

(Favorite fish shared attributes: open mouth, three bubbles, stripes, one top fin, two on bottom)
1; 5; 6; 8; 9; 12; 13

Puzzle 40 (pg 46)
Toss The Dice

1. Figure C. black background with one white dot; white background with two black dots
2. Figure F. black circle; black triangle
3. Figure I. One side should show 5, 1, 3 down the left side and 6, 2, 4 down the right side. The other side should show 6, 2, 4 down the left side and 1, 3, 5 down the right side.

Puzzle 41 (pg 47)
Rah-Rah Combos

1. three more arrangements
2. six more arrangements
3. (Tower 5) B, top; A, middle; (Tower 6) A, top; B, middle; K, bottom. (Top two letters in each tower are also correct when reversed.)
4. K

Puzzle 42 (pg 48)
Puzzling Patterns # 4

1. a and g; b and d; c and e; f and h
2. 1. S, black; 2. Q, white; 3. P, grey; 4. H, grey; 5. A, black; 6. G, white; 7. X, grey; 8. B, black

Puzzle 43 (pg 49)
Mini-Mysteries

1. Emily.
2. #21.
3. Package E, from Albany

Puzzle 44 (pg 50)
Analogy Quick Thinks # 2

1. A
2.
3. J-5

Puzzle 45 (pg 51)
Pizza Portions

BONUS: 7

Puzzle 46 (pg 52)
Prizewinning Numbers

1. *A* numbers share these attributes: even number, digit in tens place is twice the digit in hundreds place, sum of all digits is 21.

   C Numbers share these attributes: Each digit is less than the digit to its left; all are 4-digit numbers, in each, the tens and hundreds place digits total 11.

   A number cannot be in both *A* and *C* because for numbers in *A* the tens digit is twice the hundreds digit, and the size of numbers in *C* must decline as the e moves to the right.

2. Students may find several other numbers, such as 9135; 35,307; 53,307; or 10,539.

Puzzle 47 (pg 53)
Predictable Patterns

1. The first two digits added make the last digit(s). The middle letters are the alphabetical equivalents of the first two digits and the last digit. For example; 5 + 7 = 12, fifth letter of the alphabet is E, seventh is G, the twelfth is L, therefore the license plate reads 57EGL12.
   Answer a. 28BHJ10
   b. 45DEI9
   c. 62FBH8

2. Vanity plates (substitute with the alphabetical letter or number that comes before those in the license plates):
   d. 4 ever;
   e. U R A Q T
   f. good 4 u.

3. g. MA31; h. JU30. (Letters represent months, numbers are the last day of the month.)

4. i.

5. j. forgive and forget;
   k. one foot in the grave;
   l. Anyone for tennis?

Puzzle 48 (pg 54)
Cipher Words

1. d – o – t – h – o – g = hot dog
2. s – h – o – c – a – n = nachos
3. a – u – d – e – n – s = sundae
4. t – o – o – p – a – t = potato
5. c – i – q – h – u – e = quiche
6. b – e – d – r – u – p = burped

   Bonus: Check to see that student has written the code correctly to spell any four-letter word (*food*, for example) that relates to the list of words.

Puzzle 49 (pg 55)
Book Bewilderment

Attributes of librarian Higgle's books: main titles have three words; author's name or title always includes an abbreviation; spines always contain some numerals.

Books to sort: 1. B   4. B
              2. A   5. A
              3. A   6. A.

Puzzle 50 (pg 56)
A Tribute To Flags

Students must draw a flag with three simple attributes:
   two-colored background divided by one line;
   three similar geometrical shapes;
   three similar pictures.

Otherwise, size, color and placement do not matter.

Puzzle 51 (pg 57)
Squared!

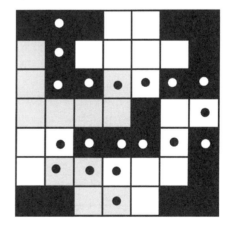

Puzzle 52 (pg 58)
Eighth Grade Musical!

Characters are:

   Moogie on keyboards

   Stringbean on drums

   Geneva on bass

   J.J. Hipp on electric guitar

   Crystal on saxophone